D1451437

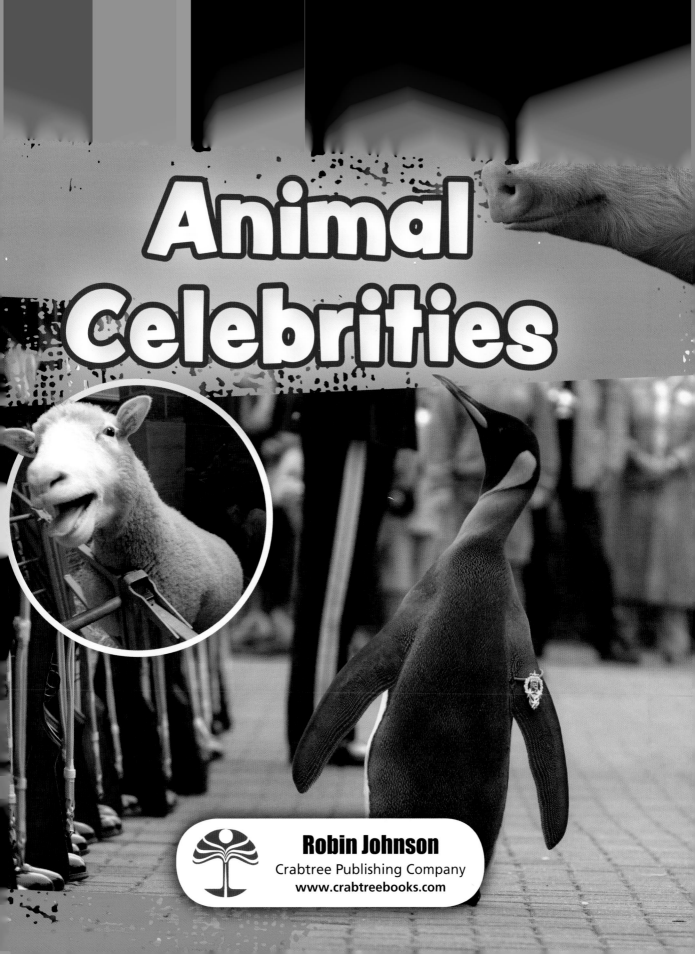

Animal Celebrities

Robin Johnson

Crabtree Publishing Company

www.crabtreebooks.com

CRABTREE
PUBLISHING COMPANY
WWW.CRABTREEBOOKS.COM

Author:
  Robin Johnson
Editorial director:
  Kathy Middleton
Editor:
  Sonya Newland
Proofreaders:
  Izzi Howell, Crystal Sikkens
Graphic design:
  Clare Nicholas
Image research:
  Robin Johnson
Production coordinator and prepress:
  Tammy McGarr
Print coordinator:
  Katherine Berti

Images:
Alamy: 10 (Newscom), 20–21, 21t (Xinhua), 25t (ITAR-TASS News Agency); Getty Images: 4 (Bruce Glikas), 6, 7b (Micke Sebastien), 9b (Odd Andersen), 11t (Toru Yamanaka), 14 (The Asahi Shimbun), 15t (Getty Images), 15b (Colin McPherson), 16, 17t (Patrik Stollarz), 18, 19t, 19b (Bernard Weil), 21b (EZEQUIEL BECERRA), 22, 22–23 (Jeff Swensen), 24 (Sovfoto), 26 (AFP), 27t (EDJones/AFP), 29b (FOX); iStock: 13t (steve_is_on_holiday), 13m (StHelena); Ministry of Defence: 27b (contains public sector information licensed under the Open Government Licence v3.0.); Shutterstock: 5 (Kathy Hutchins), 8, 9t (360b), 11b (Opasbbb), 12 (Darrin Henry), 17b (Vladimir Melnik), 23t (Celadris), 23b (karenfoleyphotography), 25b (Mitrofanov Alexander), 28 (Jaguar PS), 29t (Jstone); Wikimedia: 7t (Pilleybianchi), 13b.

Library and Archives Canada Cataloguing in Publication

Title: Animal celebrities / Robin Johnson.
Names: Johnson, Robin (Robin R.), author.
Description: Series statement: Astonishing animals |
  Includes bibliographical references and index.
Identifiers: Canadiana (print) 20200155083 | Canadiana (ebook) 20200155091 |
  ISBN 9780778769163 (hardcover) |
  ISBN 9780778769347 (softcover) |
  ISBN 9781427124340 (HTML)
Subjects: LCSH: Famous animals—Anecdotes—Juvenile literature. |
  LCSH: Famous animals—Miscellanea—Juvenile literature. |
  LCSH: Human-animal relationships—Juvenile literature.
Classification: LCC QL793 .J64 2020 | DDC j591.02—dc23

Library of Congress Cataloging-in-Publication Data

Names: Johnson, Robin (Robin R.), author.
Title: Animal celebrities / Robin Johnson.
Description: New York, New York : Crabtree Publishing Company, [2020] |
  Series: Astonishing animals | Includes index.
Identifiers: LCCN 2019053187 (print) | LCCN 2019053188 (ebook) |
  ISBN 9780778769163 (hardcover) |
  ISBN 9780778769347 (paperback) |
  ISBN 9781427124340 (ebook)
Subjects: LCSH: Animals--Anecdotes--Juvenile literature. |
  Animals--Miscellanea--Juvenile literature. | Human-animal
  relationships--Juvenile literature.
Classification: LCC QL793 .J64 2020 (print) | LCC QL793 (ebook) |
  DDC 591.02--dc23
LC record available at https://lccn.loc.gov/2019053187
LC ebook record available at https://lccn.loc.gov/2019053188

Crabtree Publishing Company

www.crabtreebooks.com   1-800-387-7650

Printed in the U.S.A./022020/CG20200102

Published in Canada
Crabtree Publishing
616 Welland Ave.
St. Catharines, Ontario
L2M 5V6

Published in the United States
Crabtree Publishing
PMB 59051
350 Fifth Avenue, 59th Floor
New York, New York 10118

Published in the United Kingdom
Crabtree Publishing
Maritime House
Basin Road North, Hove
BN41 1WR

Published in Australia
Crabtree Publishing
Unit 3 – 5 Currumbin Court
Capalaba
QLD 4157

Table of contents

Famous furballs

What makes an animal famous? Some animals are known for doing amazing tricks or having super skills. Others are hard workers that do important jobs, or are heroes that help scientists. And some animals are famous for being cute, smart, big, old—or even grumpy!

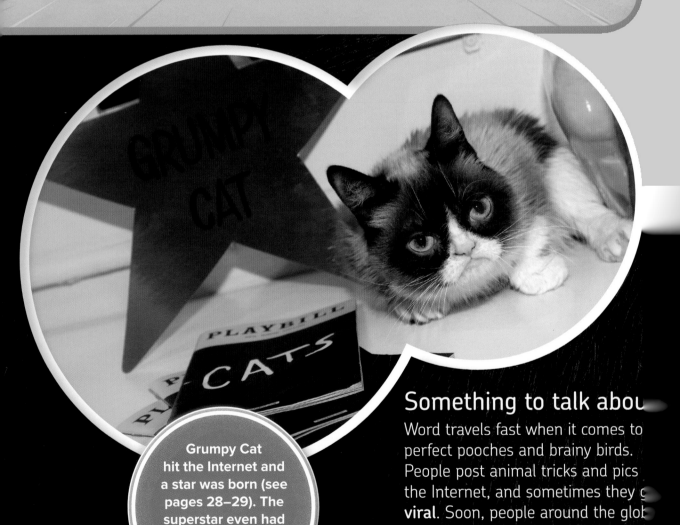

Grumpy Cat hit the Internet and a star was born (see pages 28–29). The superstar even had a guest spot in a Broadway play called *Cats*!

Something to talk abou

Word travels fast when it comes to perfect pooches and brainy birds. People post animal tricks and pics the Internet, and sometimes they g **viral**. Soon, people around the glob are sharing tales about the animals The pets become **celebrities** with millions of fans and followers

Creature comforts

Animal celebrities live a life of **luxury**. They are often loved by people everywhere, and treated like royalty at home. Fans gather to take pictures and give them treats and toys. The pampered pets have everything they could ask for—if they could talk!

Jiff the Pomeranian (Jiffpom) became famous for walking on two paws faster than any other dog. Then he became even more famous for his huge social media following!

Chaser the dog

Have you ever heard the saying "You can't teach an old dog new tricks?" Well, anyone who says that has obviously never met Chaser!

Chaser's owner showed her an object and repeated its name up to 40 times. Then he hid the object and asked Chaser to find it. He continued to repeat its name while Chaser looked.

Toy story

Chaser was a border collie—a smart **breed** of dog often trained to **herd** sheep. But Chaser's owner trained her to do something else instead. He spent hours each day teaching Chaser the names of toys. Chaser worked like a dog learning to fetch the toys by name. And her hard work paid off!

6

Move over, Rover!

Chaser became known as the smartest dog in the world. She could identify more than one thousand different toys by name! That's the largest list of spoken words learned by any known animal. Chaser was featured in magazines, newspapers, and books. She traveled the globe to show her skills on TV shows—and always bowwow-wowed the audience!

FACT FILE

Born: April 28, 2004

Lived in: Spartanburg, outh Carolina

Hobbies: Fetching toys and herding other family pets

Died: July 23, 2019

Chaser had a huge toy collection, including 800 stuffed animals, 116 balls, and 26 Frisbees!

Chaser's favorite toy was a small, blue, rubber ball.

Wow!

Chaser learned two new words each day! Young children learn about 10 words a day.

Knut the polar bear

What happens when a polar bear the size of a snowball is raised by people? He melts hearts all around the world!

Bear necessities

Knut was a polar bear born at a zoo in Germany. Like all newborn polar bear cubs, he could not see or hear, and he had very little fur. Knut depended on his mother for his survival. But polar bears may **abandon** their cubs if they think they are too weak to survive. This happened to Knut, so zookeepers began feeding and taking care of him.

Thomas even played his guitar to help Knut fall asleep—and then slept beside the bear's cage at night!

A zookeeper named Thomas took special care of Knut. He fed the cub with a baby bottle, gave him baths, and played with him every day.

Cute Knut

News of the abandoned polar bear spread quickly around the world. The little bear cub became a huge celebrity! Fans crowded into the zoo to see the cute cub in action. They bought polar bear toys and books, and sang songs about him. Knut had his own TV show, a movie about his life—and almost too much love to bear.

Hundreds of fans gathered at the zoo each day to watch Knut play. They bought stuffed animals and other toys there.

Knut grew too big for his zookeeper to handle—but not too big for toys!

The little bear cub grew quickly, and so did his popularity!

Wow!

Hundreds of children attended Knut's first birthday party! The party was shown live on TV in Germany.

Tama the cat

Have you ever heard of a superhero who could stop a train? What about a superstar cat who could stop a train station—from closing, that is.

Hello kitty!

A little train station in Japan was about to be closed for good. It was not busy enough to stay open. Then, someone had a purr-fect idea. They put a local cat named Tama in charge of the station! Tama's duties as stationmaster were to wear a hat and badge...and to take cat naps. She had her own office—with a litter box, of course—and was paid in cat food.

Fans gathered to admire the famous feline.

Tama the cat always dressed for success as stationmaster.

Train to Tama

News of the little cat boss spread faster than a speeding train. Soon people from Japan and the rest of the world were taking the train to Tama. Fans took photos with the cute cat, and bought Tama-shaped toys, mugs, and candy. The train station was a runaway success, earning more than enough money to keep it open for good. And it was all thanks to a cool cat in a little black hat!

FACT FILE

Born: April 29, 1999

Lived at: Kishi Station in Japan

Job title: Stationmaster

Died: June 22, 2015

Trains on the Tama railway line are designed to look like the celebrity cat.

Inside the station, fans can visit the Tama Museum and eat cat treats at the Tama Café.

Wow!

When Tama died, more than 3,000 people came to her funeral. They left flowers and cans of tuna for her at the station.

People still hang decorations at a **shrine** for Tama that was made at the station.

和歌山 WAKAYAMA

ワンマン

TAMA DENSHA

TAMADENSHA 2275

たま電車

TAMA

WAKAYAMA ELECTRIC RAILWAY

和歌山電鐵貴志川線

Jonathan the tortoise

Terrific old-timer tortoise

How does the oldest land animal in the world move? As slow as a tortoise!

Slow and steady

A tortoise is similar to a turtle, but lives on land instead of water. Most giant tortoises have a **lifespan** of 150 years. But a famous Seychelles giant tortoise named Jonathan has outlived them all. He is about 187 years old!

In his old age, Jonathan has lost his sight and sense of smell, but his hearing is excellent.

Jonathan enjoys spending time with people. Like all celebrities, he has many fans!

Jonathan responds when his name is called, and is fascinated by the sounds of tennis.

The test of time

Jonathan has faced many dangers in his long life. In the early 1800s, people hunted giant tortoises for food. In fact, so many were killed that scientists thought Jonathan's species was **extinct**! Today, there are only around 80 of these tortoises left in the world. The most famous is old Jonathan—a gentle giant who has stood the test of time.

Jonathan is kept in **captivity** in style. He lives on the grounds of a grand home in Saint Helena. The governor of the island lives there!

Visitors to the island can tour the home and grounds. They can see Jonathan and three other giant tortoises—Emma, David, and Fred—who live there.

Wow!

Jonathan hatched before the invention of the light bulb, telephone, and car!

This photo of Jonathan was taken in 1886, more than 130 years ago!

13

Dolly the sheep

Superstar sheep made by scientists

A sheep named Dolly was a born celebrity. But she was not born like any other sheep before her!

Seeing double

Dolly was created in a science lab as part of an experiment. Scientists wanted to see if they could clone a living **mammal**. That means they wanted to grow an exact copy of it. Scientists combined a tiny **cell** from one sheep with a cell from another. The cells grew inside a third sheep—and the cloned lamb was born.

Dolly got her name from another celebrity—a country singer named Dolly Parton.

Her birth was so important, Dolly even appeared on a postage stamp.

Dolly later gave birth to six healthy lambs of her own.

A star is born

Cloning from an adult mammal had never been done successfully before. People around the world wanted to know all about Dolly and how she was made. Dolly looked and acted like any other sheep, but her birth was a breakthrough in science. Not baaaad for a little lamb!

Reporters flocked to the animal research center to see the little lamb made in a lab.

Dolly got people around the world talking about cloning and other parts of science.

Scientists at the center received 3,000 phone calls the week Dolly was revealed to the world!

Wow!

It took scientists 276 tries before their experiment worked and Dolly was born!

15

Paul the octopus

Who had eight legs and could **predict** soccer games? A celebrity octopus named Paul!

Paul lived in a glass tank decorated with rocks and soccer balls.

Paul opened a box in his tank to get a treat—and to pick a winning soccer team.

Hundreds of reporters and fans crowded around Paul's tank to see which team he would choose.

World Cup octopus

Octopuses are smart animals. They love to solve problems and figure things out. In 2010, an octopus named Paul, who lived in an aquarium in Germany, amazed the sports world. His keepers wanted to see if he could pick the winners in the wildly popular World Cup soccer competition. So they put flags of the competing countries on food boxes and placed them inside his tank. Whichever box Paul opened first would be his pick to win the next soccer game.

Eight-legged legend

Could a sea creature really see into the future? Somehow, Paul predicted the winner in all seven of Germany's games, plus the final between Spain and the Netherlands! He became an unlikely star of the World Cup. Fans waited eagerly for Paul to choose their team—and joked that they would cook and eat him when he did not!

FACT FILE

Born: 2008

Lived in: An aquarium in Germany

Hobby: Taking the lids off jars

Died: October 26, 2010

Despite threats from unhappy fans, Paul lived for more than two years— a full life for an octopus.

Soccer fans filled stadiums in South Africa to watch the 2010 World Cup—and see if Paul's predictions would come true.

Wow!

There was a one in 256 chance that Paul would predict the winners in all of Germany's games!

Fans wore octopus gear and sang silly songs about Paul. One song claimed that the octopus had "magical" tentacles.

Esther the Wonder Pig

What happens when your mini pet becomes the size of a mini bus? You get Esther the Wonder Pig!

This little piggy stayed home

As an adorable little piglet, Esther could fit in the palm of your hand. She weighed just 4 pounds (1.8 kg) when a Canadian couple adopted her and brought her home. They were told that Esther was a micro pig who would grow to be the size of a dog. But their new pet pigged out and grew like any other pig would do. Two years later, Esther was a full-sized sow!

Massive micro pig

Esther eventually grew to weigh a whopping 600 pounds (272 kg).

Esther learned how to open kitchen cupboards and fridge doors to get food.

Happily ever Esther

As Esther grew, so did her popularity. Her owners posted pictures of the pretty pig playing, eating, and sleeping. Esther hammed it up in funny wigs and outfits, and became a social media darling. With help from Esther's followers and fans, her owners raised enough money to open a farm **sanctuary**. Today, the superstar swine lives there with other rescued farm animals—and is as happy as a pig in mud!

Esther is a pampered pig. She has her own bedroom in the farmhouse—complete with a comfy king-sized bed!

Esther enjoys playing tag and tug-of-war with her owners.

Fans can buy autographed photos of Esther stamped with her hoofprint.

The sanctuary is called Happily Ever Esther Farm Sanctuary.

Wow! Esther the Wonder Pig has nearly 1.5 million Facebook followers!

19

Grecia the toucan

One can, two can, we all can help a toucan!
(Or at least some clever scientists can!)

Sad song

Toucans are big, brightly colored birds with large beaks. Sadly, a toucan in Costa Rica named Grecia was attacked by a group of young people and lost the top half of its beak. Without a beak, Grecia would not survive. Luckily, the injured bird was found by farm workers and rushed to an animal rescue center for help.

The poor bird tried to eat and sing, but was unable to do either.

Toucans use their beaks to get food, make nests, and defend themselves.

Grecia was named for the town in Costa Rica where it was found.

Fit the bill

News of the toucan took off. Soon, a team of scientists, doctors, dentists, and engineers from around the world were working together to make a new beak for Grecia. They used a **3D printer** to make and test many models before finding one that fit the bill. Now Grecia has a shiny new **prosthetic** beak—the first one ever made for a toucan.

FACT FILE

Hatched: Around March 2014

Lives in: Grecia, Costa Rica

Diet: Fruits, insects, lizards, and eggs

Scientists left Grecia's beak white to remind people of the abuse the bird had suffered.

It took Grecia more than a year to recover from its injuries.

Grecia can eat like a bird again—and is singing sweeter than ever.

Wow!

The celebrity bird's story inspired Costa Ricans to demand laws to protect animals from abuse.

Phil the groundhog

Snow, snow, go away.
What will Phil
predict today?

Fantastic furry forecaster

Groundhog Day

A grand event takes place on February 2 each year. People gather in a Pennsylvania town to watch a little groundhog named Punxsutawney Phil. They wait in the freezing cold for hours just to see the superstar **rodent**. Why? Because legend has it that Phil can predict when spring will come each year!

Reporters are eager to find out furry Phil's forecast.

Thousands of people gather to see Phil each year. Some wear their finest clothes. Others dress up like groundhogs!

Traditionally, a groundhog goes back inside its hole if it sees its shadow. Punxsutawney Phil is often brought out again to party with the crowd!

Spring in his step

Phil pops out of his winter den at sunrise and looks cautiously around. If he sees his shadow and goes back in his den, it means there will be six more weeks of winter. If he does not see his shadow, spring will come sooner than that. The crowd goes wild no matter what he predicts!

A groundhog has predicted the weather in Punxsutawney since 1886. There have been many different groundhogs named Phil.

Wow!

Punxsutawney Phil is so famous, he has a holiday named after him! February 2 is "Groundhog Day" in the United States and Canada. Other celebrity groundhogs predict the arrival of spring in different parts of these countries.

Phil's den is found in a small wooded area called Gobbler's Knob.

The first known celebration of Groundhog Day was in 1840 in another part of Pennsylvania, but the tradition may have begun earlier.

PUNXSUTAWNEY
PENNSYLVANIA

The groundhog is a rising star— who predicts the weather as the sun rises.

Laika the dog

Sit, stay, roll over, fetch—blast off?
Have you ever heard of a dog trained
for a space mission?

Good dog

In 1957, scientists in the **USSR** were testing space travel to make sure
it was safe for humans. They decided to send a dog into **orbit** to see how
living things were affected by being in outer space. They found a stray
dog with a gentle, **obedient** nature and began training her for the mission.
The scientists named the dog Laika, which means "Barker" in Russian.
Soon, her name would be known around the world.

Laika was
launched in the
Sputnik 2 spacecraft.
The padded cabin had
just enough room for
her to stand up
or lie down.

Laika was
believed to be part
terrier and part husky.
She was a tough little
dog who had survived
on the streets
of Moscow.

Laika was
also called Little
Curly, Little Lemon,
Little Bug, Little
Lady—and even
Muttnik.

Space dog

Laika learned to stay in a small space for days, and then weeks at a time. After months of training, she was rocketed into space. Laika survived the launch and successfully orbited Earth, giving scientists the first real information about living things in space. Sadly, the spacecraft overheated and Laika did not survive the mission. But the tale of the four-legged hero and her famous space flight will live on forever.

FACT FILE

Born: Around 1954

Lived in: Moscow, Russia

Orbited Earth in: *Sputnik 2*

Died: November 3, 1957

Laika was a superstar. Her picture was put on postage stamps and many other products.

80

OLIA · МОНГОЛ ШУУДАН

Doctors placed medical devices in Laika's body to monitor how her body worked in space.

It took months to prepare Laika for the mission, but only 103 minutes for her to orbit Earth.

Wow!

Laika was the first living creature to ever orbit Earth. Now that's something to bark about!

Nils the penguin

What do you call a king penguin who is large and in charge? Sir, of course!

Nils Olav the king penguin lives in a zoo in Scotland. The Norwegian Guardsmen come to see him when they are visiting the country.

Soldiers beat drums and fans go wild when Nils Olav waddles by.

The penguin even carries out inspections on the soldiers under his command!

Long live the king!

Nils Olav looks like any other king penguin at the zoo. He swims, eats fish, and waddles around with the other penguins. But Nils is a V.I.P.— a very important penguin. He is an **honorary** soldier in the King of Norway's royal army, the Norwegian Guard.

Flying high

Nils is the third penguin to be honored by the Norwegian army. The tradition began in 1972 when a soldier visited the zoo in Scotland and adopted a penguin. Since then, Norway has promoted penguins at the zoo as an act of friendship between the two countries. Nils has climbed the ranks and now holds the important title of Brigadier. He has also been **knighted** by the King of Norway!

FACT FILE

Lives in: A zoo in Scotland

Height: 3 feet (0.9 m)

Job: Brigadier in the Norwegian army

Known as: A born leader

Nils, of course, wore black and white to be knighted. He is the only knighted penguin in the world!

Nils's knighthood medal was carefully tied to his right flipper.

Wow!

The superstar bird's full name is Brigadier Sir Nils Olav III! He is named after two people: a former king of Norway, and the army officer who adopted the first penguin.

27

Grumpy Cat

Some animals are famous for their skills, jobs, or ranks in the army. Others are just sour pusses!

Cranky kitty

Tardar Sauce was a cat who looked like she was always frowning. Her photo was posted on social media in 2012 and went viral. She became an Internet star with the nickname Grumpy Cat. The famous feline had fans around the world—and more than eight million followers on Facebook! Soon, people were lining up to see the cranky kitty in real life.

Tardar Sauce was not really a cross cat! Her face was just shaped in a grumpy way.

It pays to be a sour puss! Tardar Sauce and her owner got rich from the cat's frowny face.

28

On the cat walk

Tardar Sauce became the most popular pet in Hollywood. She starred in a movie and had guest spots on TV shows. She made commercials and a music video. She even "wrote" bestselling books. But not even being rich, famous, and living a life of luxury could turn Tardar Sauce's frown upside down!

Tardar Sauce's face appeared on T-shirts and many other items with slogans such as "Go away" and "I hate everything."

FACT FILE

Born: April 4, 2012

Lived in: Morristown, Arizona

Hobby: Hiding behind curtains

Died: May 14, 2019

Fans were thrilled to pose with the famous cat—who never smiled for photos.

Tardar Sauce met actors, singers, and other celebrities on TV shows such as *American Idol*.

Wow!

Tardar Sauce was flown first-class to shoot a cat food commercial. She had her own king-sized hotel room, a driver, and even an assistant to brush her fur!

29

Glossary

3D printer A device used to build objects from computer drawings

abandon To leave and never return to someone or something

breed A particular type of a species of animal

captivity An enclosed space that is removed from the wild

celebrity A person or animal who is known or recognized by a lot of people

cell The basic part that makes up all living things

extinct Describing an animal or plant species that no longer exists

feline A cat

herd To gather and move a group of animals

honorary Given as a sign of respect with no duties required

knighted To be given a special honor and the title of "Sir" by a king or queen

lifespan The average length of time a type of animal lives

luxury Something fine or special that costs a lot of money

mammal A warm-blooded animal that usually gives birth to live young

obedient To do what you are told

orbit To travel around something, such as a planet or moon, in a curved path

predict To say something might or will happen in the future

prosthetic A fake part of the body made to replace a part that has been damaged or lost

rodent A small animal, such as a mouse or beaver, that has sharp front teeth

sanctuary A place that provides safety or protection

shrine A place that people visit because it is connected with someone or something important to them

USSR The Union of Soviet Socialist Republics – a country in Eastern Europe and northern Asia that existed between 1922 and 1991, which included Russia and 14 other republics

viral Spreading very quickly to many people, especially through the Internet

Find out more

Books

Beer, Julie and Harris, Michelle. *Pet Records*. National Geographic Children's Books, 2020.

Gallo, Ana. *Pets and Their Famous Humans*. Prestel Junior, 2020.

Moberg, Julia. *Historical Animals: The Dogs, Cats, Horses, Snakes, Goats, Rats, Dragons, Bears, Elephants, Rabbits and Other Creatures That Changed the World*. Charlesbridge, 2015.

Websites

cbc.ca/kidscbc2/the-feed/tama-and-nitama-the-stationmaster-cats
Hop on the Tama train to learn more about the famous cat.

edinburghzoo.org.uk/animals-and-attractions/sir-nils-olav/
Watch a march of the famous penguin at the Edinburgh Zoo's website.

guinnessworldrecords.com/news/2019/2/introducing-jonathan-the-worlds-oldest-animal-on-land-561882/
Read about the life and times of the oldest land mammal here.

Index